What Every Technical Author Needs to Know

IEEE

IEEE⊕computer society

⊕CSPress

Press Operating Committee

Chair

James W. Cortada
IBM Institute for Business Value

Board Members

Mark J. Christensen, Independent Consultant
Richard E. (Dick) Fairley, Founder and Principal Associate, Software Engineering
   Management Associates (SEMA)
Cecilia Metra, Associate Professor of Electronics, University of Bologna
Linda Shafer, former Director, Software Quality Institute, The University of Texas
   at Austin
Evan Butterfield, Director of Products and Services
Kate Guillemette, Product Development Editor, CS Press

IEEE Computer Society Products and Services

The world-renowned IEEE Computer Society publishes, promotes, and distributes a wide variety of authoritative computer science and engineering books, e-books, journals, magazines, conference proceedings, and professional education products. Visit the CS Store at www.computer.org/store to see the current catalog.

To submit questions about the books and e-books program, or to propose a new title, please e-mail books@computer.org or write to Books, IEEE Computer Society, 10662 Los Vaqueros Circle, Los Alamitos, CA 90720-1314. Telephone +1-714-821-8380.

What Every Technical Author Needs to Know

by David Alan Grier and Evan Butterfield

IEEE ReadyNotes Series

IEEE ⏀ computer society

⏀ CSPress

Page design by Monette Velasco.

ISBN-10: 0-7695-4820-2
ISBN-13: 978-0-7695-4820-3
IEEE Computer Society Order Number: P4820

About the Authors

David Alan Grier is a long-term writer for the IEEE Computer Society and has served as Vice President of Publications for the Society. He is the author of books such as *When Computers Were Human* (Princeton 2005) and *Too Soon to Tell* (Wiley-IEEE Computer Society 2009). He writes the column and blog "The Known World" for *Computer* and, among other courses, has taught composition at the George Washington University. He is currently an associate professor of International Science and Technology Policy.

Evan Butterfield is the Director of Products and Services for the IEEE Computer Society. Previously, he was Vice President of Product Development and Publisher for Kaplan Professional in Chicago and taught college composition classes at George Washington and American Universities in Washington, DC. He received his JD from DePaul University in Chicago and has an unnatural interest in copyright and intellectual property issues. He now lives in Southern California.

Acknowledgements

No honest author believes that he or she is the sole creator of a work. Many people and many ideas influence any piece of writing. They all deserve more thanks than a few simple sentences can convey.

Jennifer Carruth deserves much gratitude for her contributions. She not only outlined the ideas for this Note but also provided much of the inspiration. In her role as the staff member for the Computer Society's author misconduct process, she helped convince us that such a Note as this would be useful and that it would likely reduce the number of plagiarism cases substantially as well as improve the quality of writing for the Society's publications. Her help on this Note and in her role in the society are gratefully acknowledged.

A long list of Computer Society staff and members added their contributions, big and little, to this publication. The staff knows who they are. The members include David Bader, Alain April, and Jean-Luc Gaudiot.

And a final, special thanks to Kate Guillemette for editing this Note.

Contents

About the Authors   v

Acknowledgements   vii

Introduction   xi

Chapter I:
    Structure of a Technical Article   1
        Section 1: The Introduction and Statement of Idea   3
        Section 2: Research Framework   6
        Section 3: The Result   10
        Section 4: The Argument   12
        Section 5: The Recap or Summary or Conclusion   14
        Section 6: Other Elements of a Technical Paper   15

Chapter II:
Copyright Law: What It Is, and How Not to Break It   **19**
    Section 1: Copyright   20
    Section 2: Fair Use   23
    Section 3: The IEEE and Copyright   27

Introduction

This ReadyNote is intended to help you prepare technical ideas for publication and protect the publications that contain those ideas. Publication is a key element of science and engineering, as it not only spreads new ideas and practices but also sustains technical communities. Papers are part of a conversation in which an author builds upon the work of the past and leaves ideas that others can develop.

Technical writing has its own particular challenges. When done well, it can present papers that are a delight to read, that present new ideas, and that remind the reader of the great joys of scientific work. When done badly, papers can be tedious or misleading or even incomprehensible. In the worst cases, papers can impinge on the rights of others and claim that the authors should receive credit for work that was not their own.

To prepare a good technical article, authors need only to have a new idea and understand a few basic principles. They need to know a few basic rules of expository writing, the requirements of the type of article that they intend to write, and the standards that protect intellectual property. These topics are the basic subjects of this ReadyNote. This Note is intended for new authors, for engineers and scientists who are just starting to write and publish. Though it presents ideas that can be used by any technical writer, it is particularly targeted at those researchers and practitioners who are writing for the Computer Society and its parent organization, the Institute of Electrical and Electronic Engineers.

Though publication is an important step in disseminating ideas, it necessarily involves giving up some of the rights to those ideas. When you put a paper before the public, you are inviting others to engage your ideas and utilize them. One of the means that we use to protect ideas is to copyright the paper. Copyrighting not only establishes your claim to the paper itself but also to the ideas behind the paper. As different publishers deal with the issue of copyrighting in different ways, this ReadyNote provides an overview of the copyright process and describes the various ways of protecting the rights to your work.[1]

There are many ways of approaching the problem of writing technical articles, publishing information, and protecting ideas. This ReadyNote presents a few strategies that are simple to use and have been successful in the past.

---

1   We need to note that this ReadyNote does not deal with the issue of patenting, which is a second way of
protecting intellectual property.

Chapter I:
Structure of a Technical Article

O f course, the technical world has no universal standard for the structure of technical articles. Search the IEEE digital library as much as you wish, and you will find no approved standard for technical papers. Yet, if you skim through the contents of that same digital library, you will find a remarkable uniformity in the structure of the papers it contains. The papers will have similar divisions, similar logic, and similar language. They contain these similarities because they are all attempting to solve a common problem, the problem of introducing a new technical idea, connecting that idea to an existing body of knowledge, and arguing that this idea is good and valid.

As you prepare to write a new article, you should focus on these key tasks of the technical paper: the goals of introducing your idea, connecting it to the existing literature, and arguing the validity of your research. By focusing on these tasks, you will see that the guidelines in this ReadyNote are not arbitrary rules but principles that are designed to help you meet your goals. If we leave you with one idea, we would tell you that you can ignore any of our guidelines if they get in the way of a clear, well-written paper. At the same time, we would tell you that these guidelines have worked well for generations of scientists and engineers and would warn you that you ignore them at your peril.

We will start with the basic structure for the technical paper, the common form that will help you communicate your idea. By starting at this point, we are skipping the fundamental topics of English grammar, syntax, and composition. These are important topics, but they are already well treated by the composition literature.

The common structure for scientific and technical papers has five basic sections. Those sections are:

The Introduction and Scientific or Technical Context
The Research Framework
The Research Result
The Research Argument
The Recap

In addition to the five main sections, articles have a few additional elements. These include an abstract, a bibliography, an author biography, and a set of keywords. All of these sections and elements will be discussed below.

To help clarify the structure of a paper, we will use an example drawn from the IEEE-CS Digital Library. This example, "Self-Stabilizing Clustering of Tree Networks," is taken from the *IEEE Transactions on Computers*.[1] It has the five basic sections of an argumentative paper, though it uses slightly different labels. In it, the five sections are called:

1. Introduction
2. Preliminaries
3. Algorithm
4. Correctness
5. Conclusion

These five sections in this paper serve the same purpose as the five sections that we have identified above. The first section, called "Introduction" in the example, introduces the reader to the basic problem that the paper attempts to solve. The second section, "Preliminaries," establishes the scientific context, explains the notation, and defines the basic concepts. The third, "Algorithm" in the example, describes the basis result. The fourth, "Correctness," provides the argument for the validity of the result. In this case, the argument is a structured like a mathematical proof. In many articles the argumentative section will be an empirical or statistical study. The last section, labeled "Conclusion" in the example, summarizes the work, assesses the accomplishment, and suggests further research.

---

1   M.H. Karaata, "Self-stabilizing clustering of tree networks," *IEEE Transactions on Computers*, vol. 55, no. 4, 2006, pp. 416–427.

To understand better the goals of each of these sections, we will examine them one by one. As we consider each section, we will review the goals of each part and suggest strategies for accomplishing those goals.

Section 1: The Introduction and Statement of Idea

The structure of the Introduction to a technical paper is determined by one of the fundamental properties of good writing. Good technical writing presents ideas in a way that can be understood easily. In particular, it should allow readers to start at the beginning and read straight to the end without backtracking. Few things are more frustrating than a paper that forces a reader to cycle back through the paragraphs to find where ideas were introduced, what terms mean, how conclusions were reached. If readers can start at the beginning of the paper, read straight to the end without returning to prior sections, and feel that they understand the idea, then they will feel generally that the paper is well written.

Using this goal as our guide, we can see that the Introduction of a technical paper should provide the reader with a foundation of familiar material before moving into new ideas. From that foundation, it should establish the field of research, the kind of problem that you are solving, and the nature of the solution. After completing the introduction, readers should know the subject of the paper and the kinds of arguments that it will be using.

The centerpiece of any article is the result, the new idea that you are presenting to the world. You want to present that idea to your readers as quickly as you can so that they will understand the topic of the paper and the kind of work that you are doing. However, if you start a paper merely by stating your idea, you risk overwhelming and confusing those readers.

Because you are presenting a technical idea, especially if the idea is expressed in symbols, you usually need to present it in a context that will give some meaning to the idea and also suggest the nature of the research that you have done. At the same time, if the paper begins with a lengthy discussion of the context, any readers can find themselves bored and confused. They won't know what problem you are trying to solve and what benefits your solution may have. Therefore, the first section of the paper needs to establish three things:

The problem that you are trying to solve;
The context in which you have solved it;
The benefits of your solution.

Therefore, the first part of the paper needs to be an introduction, something that introduces your readers to the problem that you are solving, the context in which you are solving it, and the benefits of your solution. As we will see, the introduction may not give a complete statement of the idea, and that statement may be a little incomplete or vague. It is more important to engage the reader with your introduction than to present every element of your idea.

Identification of the Field. Many aspects of the field will be obvious from the periodical containing the article. One does not expect to see a software paper in a transaction devoted to hardware problems, for example. However, the paper should quickly establish the exact nature of the field that it is exploring. Many papers can identify the field in a single sentence. If we look at the first sentence of the Karaata paper, we find:

> Clustering is defined as the covering of the nodes of a network by subtrees such that the intersection of any two subtrees is at most a single node and the difference between the sizes of the largest and the smallest clusters is minimal.

This paragraph identifies the subject of the paper as clustering within a network. The second sentence helps clarify the subject of the paper as identifying or building clusters within networks:

> Partitioning of computer networks into clusters is a fundamental problem with many applications where each cluster is a disjoint subset of the nodes and the links of the network that share the usage of a set of resources and/ or services.

Admittedly, these two sentences are fairly abstract and require that the reader have some technical knowledge in order to understand the full nature of the field. However, this paper attempts to further clarify the field by giving some examples in the final sentences of the first paragraph. By the end of that paragraph, the reader should understand the topic of the paper or have recognized that the paper is on a topic that is not of interest to them.

Discussion of the Context. Normally, the identification of the field requires only one paragraph. In some cases it can be done in a single sentence. However, scholarly papers need to put themselves in a well-defined context. Technical papers never stand alone. They are always part of a conversation that involves other papers and other authors. As soon as it has identified the field, the scholarly paper needs to establish that conversation. It needs to identify the other papers that are part of this conversation.

The section that discusses the context of the problem need not be large, but it does need to cover all of the relevant literature. If it fails to discuss key papers, it can easily be identified as weak or incomplete by editors or referees. At the same time, if it includes far too many papers, it can give the impression that the authors do not know what they are doing.

We cannot give a fixed rule for the number of papers that should be included in this section. Sometimes, it might be just a handful, especially if the field is new or has a good survey article. In other fields, especially those that are very active, this part of the paper might discuss 30 or even 40 other references.

Typically, this part of the paper can be divided into 4 subsections, which we will represent as four sentences. That structure starts by giving the foundation of the field. It then gives a set of papers that present the conventional solution to the problem that is the subject of the current manuscript. In presenting those papers, it characterizes the strengths and drawbacks of their approach.

Next, it presents an alternative approach to the problem and describes the papers that have approached the topic with that alternative method. Finally, this section presents papers that are most similar to the approach that the author(s) will give in the sections that follow. A quick example of such a section would look like:

> The foundations of this problem were described in [16],[22], and [8]. In particular, [8] showed how this problem was related to a much larger class of network problems. The conventional approach to this problem, which involves a sequential algorithm, was presented in [2],[5],[6], and [7]. All of these have $O(n^2)$ complexity, though [7] showed that special cases can be $O(nlog(n))$. The parallel approach is discussed in [9],[1], and [11]. It has only an $O(n)$ complexity but requires $O(n^4)$ storage. Our approach, the semi-parallel algorithm, was discussed by ourselves in [8] and by other authors in [13]. Our result showed that the algorithm had $O(n)$ complexity and required no more than $O(n^3)$. The result in [13] reduced the memory usage to $O(n^{2.5})$. In this paper, we reduce the memory requirements to $O(n^{2.1})$.

In the example from Karaata, the discussion of the field occupies paragraph 2 and part of paragraph 3. It begins with the sentence: "The problem of clustering is closely related to the graph theoretic problem."

Proposal for a Solution. The Introduction has to give the reader some idea about the new idea, the solution, that will be discussed in the remaining part of the paper. Sometimes it will present a full explanation of the solution. In other situations, it will give only a sketch. In all cases, it should state the benefits of the solution, why the result it presents is better than prior results, how it handles new problems, or how it generalizes an older approach. If readers have no appreciation of the result or the benefits of that result, then they have no incentive to continue reading.

The example paper by Karaata proposes its solution in paragraph 4. It begins, "In this paper, we present a simple self-stabilizing distributed algorithm for clustering of rooted directed trees."

A Few Additional Guidelines. In writing this section, many writers are tempted to include every fact that is relevant to the research and feel that they have not done a good job unless they have included every idea that is relevant to the work. Resist that temptation. It is not leading you to a good paper. It will encourage you to write an introduction that is so filled with ideas that it will overwhelm the reader.

In writing, details are little spotlights; they draw attention to themselves. For example, compare the sentence

> This method solves the problem of discontinuity that has traditionally been associated with the classical theory of information.

with the sentence

> This method solves the problem of discontinuity, as identified by Schroeder and Van Pelt [6], in normal operating conditions after initial transients have been removed [7], though the system need not be in steady state, as is required by the Levy distributions of Brown and Bowen [11], but can indeed be in the secondary state of operation or approaching the secondary state, as described by Blaze [32], which are normally considered in the literature to be the requirements of the classical theory of information.

The second statement is true and may have a place in the paper. However, it is filled with details. In trying to make sense of the sentence, readers encounter three or four details and have few clues that help them understand which of these details are important. They are blinded by having too many little searchlights in the sentence. While all of the ideas may be important for the paper, they may be better presented in the second section, which describes the background for the research. In the introduction, you want simple sentences that help the reader understand the basic nature of the result. A good approach would be to combine the first sentence with the references of the second.

> This method solves the problem of discontinuity that has traditionally been associated with the classical theory of information [6][7][11][32].

You will have a place to describe the details in the second section.

Section 2: Research Framework

Once you have engaged the reader with your introduction, you then need to explain your idea. However, before you can make that explanation, you need to establish the context for the idea. For technical ideas, you cannot fully explain the idea without identifying the context, the prior literature, and even the major results of the context. Different parts of the field use common terms in slightly different ways. Different parts also emphasize different aspects of research and value different kinds of results. Therefore, the second section needs to explain the context for your work. Sometimes this section is so complicated that it will need to be divided into two or more parts. Still, all of these parts will be explaining the context of your work.

Unlike the introduction, the second section of the paper can be highly detailed. The purpose of this section is to give the reader enough information to be able to reproduce your research exactly as you did it. The communities of scientists and engineers subscribe to the fundamental idea that anyone should be able to reproduce any experiment or any form of reasoning. Of course, in practice, it is not always possible to reproduce the work behind the paper. Still, the section should give enough information about the experiment that all readers would feel confident that the research was done well.

As with the Introduction section, this section is usually most successful when it is written in a straightforward, linear way, a way that allows the reader to start at the beginning and move to the end without having to backtrack. This goal suggests that the section should have all the material that is needed to understand the research and nothing more, and that every idea should build upon its predecessor.

Most authors find it easiest to write this second section by asking a series of questions that move backwards. The first of these questions is, "What results does the reader need to understand to be able to appreciate my result?" Moving backwards, we ask, "What aspects of the research or experimental design does the reader need to know to understand those results?" "What is the mathematical model underlying that research?" "What is the notation underlying that model?" "What are the basic concepts behind the notation?" Finally, we ask, "How does all this material fit into the basic problem or idea that motivated the research?"

Once you have answered these questions, you can take your answers, reverse the order of the questions, and write a section that is structured in the following way:

Typical problem or motivating problem
Basic concepts
Notation
Mathematical Model
Research Design
Preliminary Results or Lemmas

The exact information within this section will depend on the nature of the research. Theoretical research has one approach to research, while empirical work has a very different approach. In theoretical work, it is very important to clarify the notation and carefully describe the underlying conceptual model, including all the fundamental assumptions. For empirical work, you often have to describe not only the model and notation but also the equipment and experimental design.

In large research projects, a single machine or set of equipment might generate many papers. In such settings, it is often tempting to take text from one paper that describes the experiment and copy it to a second one. While such copying does not reproduce any unique ideas, it is often caught by the anti-plagiarism systems that are used by journals and conferences. Depending on the policies of the periodical, such copying can be considered self-plagiarism and prevent the paper from being published. It is always better to be honest about such copying. You could remind the reader of the first publication concerning that experiment and then repeat the text from that paper in quotes:

> The experimental setup was described in our former paper [3]. In that
> paper, we said, "The experiment consists of …."

Some editors will say that you don't need to describe the experiment a second time. Some will tell you that you don't need to put it in quotes. Let the editor tell you what to do. However, you have protected yourself from the charge that you have borrowed text form an earlier paper by clearly stating the source of the sentences.

If you look at the example, Karaata, you will find that the research framework is described in section 2. In that section, Karaata uses three subsections to describe how he conducted his research. The first subsection (2.1) is a typical problem that is supposed to motivate the need for his research. Note that it has several references to earlier papers that describe similar research.

The second subsection (2.2) describes the notation that he will use. Unless your notation is highly standardized, you will need at least a few sentences to describe what your symbols will mean. The third subsection (2.3) describes the symbolic model that will be the basis for the author's reasoning.

A Few Additional Guidelines. Many, many guides to technical writing argue that scientific articles should be written in the passive voice, a grammatical form that hides the actor. An example of a passive sentence would be:

> *Two software modules were produced.*

Most of these guides are recommending the passive voice because they are trying to keep writers from inserting themselves into the paper. Through a long history, scientists and engineers have avoided putting themselves into their writing as a way making it easier for the paper to be criticized by others. Most scientists find it easiest to criticize a result when they are focusing on the idea rather than the author. Hence, they would avoid sentences such as:

> *I wrote two software modules.*

In general, it is good to avoid making yourself a character in your scientific paper. Since the scientific community now expects that articles are written without reference to the author, it concludes that any article written in a personal voice, with the use of words like "I," "we," "mine," and "ours," must not be written by skilled researchers.

At the same time, the passive voice can produce complicated and confusing paragraphs because it hides the thing or object that is doing the action. In English grammar, we normally structure sentences in the form:

<subject> <verb> <object>

which is equivalent to less technical description

<actor> <action> <thing that is being acted upon>

When you write in the passive voice, you flip the order and get sentences of the form:

<Thing being acted upon> <action>

In this form, the actor gets lost. Consider the following sentence:

The data was transformed.

In it, the data is the thing being acted upon but what is doing the action? The reader can only guess. The context may make it clear, or it may not.

A stronger English sentence would be:

The new algorithm transformed the data.

as this form clearly identifies the object that is doing the action, the action itself, and the object being acted upon.

Now, scientific writers often have trouble finding a good actor, especially when they are trying to keep themselves out of the paper and avoid words like "I" or "me." They end up writing sentences like:

A new communications system was produced.

In some cases, these sentences will work well and explain the idea perfectly. In many cases, however, they can get confusing. You can often avoid such confusion by recognizing that the research itself can be used as an actor. You could rewrite the above example to say:

Our research produced a new communications system.

Such a strategy can often produce a clearer description of a research plan. Instead of relying on the passive voice, you can identify aspects of the research that act on special parts of your work. Consider the following paragraph:

> The metadata from graphical or photographic input is created. Key objects from the input are identified. Object spatial relationships are established. Information is held in a relational database. Conclusions are drawn from that database and are corrected. Highly successful it was in indentifying wrong methods.

This text hides much about the very nature of the work. As we read the paragraph, we cannot be certain if the results are produced by the research, the system, the algorithm, or the user. To clarify those points, we start by identifying the research as producing the results and move through the paragraph. At each sentence, we ask, "Who or what is doing the work?" and name a subject or actor for that entity.

> The research produced a system that created metadata from graphical or photographic input. That system identified key objects in the input and established spatial relationships among those objects. It then created a relational database to hold information. The last part of the research was a special query algorithm that could reason from the database and correct its own mistakes.

Section 3: The Result

Typically, scientific and technical papers present the fullest statement of their fundamental result in the middle of the paper, in the third section. At this point, the reader understands the problem that the paper is attempting to solve, the concepts that are needed to comprehend the result, and the notation that describes the result. In most cases, the author can focus on the fundamental accomplishment of the research and not worry about anything else.

It is fairly easy to see why the final result has to appear at this point. Before you can accurately describe the result, the author has to describe the intellectual context and the research framework. Without that work, the result will make little sense. While the full statement can occur a little later in the paper, it often needs to be followed by a logical proof, or derivation, or data analysis to show its validity. Therefore, the most complete description of technical research is usually found in the middle of the paper.

Fundamentally, this section contains descriptions, words that describe the actions or ideas of the research. In writing these descriptions, authors should create a description that makes sense to the reader, one that helps the reader understand how the result works. Authors find this task easiest if they organize their thoughts around a structure or storyline. Often that structure is driven by time. Many computer science articles are organized around the transformation of data over time. "The algorithm takes data in this form as input," the story begins. "Then it transforms it into that form and finally normalizes it to a final form that it produces as output."

There are other ways to describe the result of research. One of these is a hierarchical structure. The author begins by giving an overview of the most important result and then moves into more and more specialized cases. An equally valid strategy is for the author to begin with a simple result that may not be complete and then move to more and more complex versions until they reach the final result.

In the example paper, Karaata describes his fundamental results in section 3. He organizes that section in a way that moves from the simplest result to the most complex, which is a useful approach. In subsection 3.1, he describes the fundamental results that form the basis for the algorithm. In subsection 3.2, he presents a simple version of the algorithm, one that is based on the preliminary results but is not optimal. In the final subsection, 3.3, he describes the complete, optimal algorithm.

In writing this paper, Karaata might have decided to put part 3.1 into section 2, as it describes the approach to the problem. However, the ideas in it form the basis for the two algorithms and hence can be placed equally well in section 3. This is the kind of decision authors are free to make as they write their papers.

A Few Additional Guidelines. In writing this section, authors need to think about communicating the issues that are important or new in their research. If they don't help the reader identify the most important aspects of the work, then the reader will be free to guess which elements are important and which are not. In such a case, the reader may or may not guess correctly.

In this section, authors need to present all the details of their work. They cannot overlook details here, as they might have done in the introduction section, because those details were not that important. If the author does not put the details in this section, then they will never be communicated to others. However, if the author just lists the results with all their details, then readers may not be able to determine what is important and what is not. As was mentioned in the first part of this ReadyNote, details are little spotlights. If you put too many of them in a section, they will blind the reader.

There are five basic ways of presenting results. These are:

- Temporally: Telling a story that starts at the beginning of the process and ends at the end (this is sometimes called the "Judsonian Rule," after the great American science writer, Horace Judson);
- Importance: Start with the most important result and then move to the least important;
- Specificity: Start with the most specific result and then move to the most general;

- Generality: Start with a general result and then show how it applies to specific cases;
- Simplicty: Start with the simplest result—the result that is easiest to explain. Then move to more complex results.

These rules are often combined in a single paper. The results section may follow the simplicity rule in its structure. It may start with the most elementary results and move to the most complex. However, each result may then be presented in a temporal way.

At all costs, authors should avoid the "And then." This is a mere list of results that does nothing to help readers understand which things are important and which are not. It starts by saying "This is one result." It then continues by saying "And then here is another, and then here is another," until it has presented all its ideas.

Section 4: The Argument

In all scholarly articles, section 3, the result, is the most important. If there is no result, then there is no importance. However, the next section, the argument, is vital. If the argument is no good, the paper will never be published, even if the result is good, valid, and interesting.

In scientific and engineering papers, the arguments tend to follow one of two different styles. The first is the mathematical proof, which establishes the validity of the result through logic and symbolic manipulation. The second is the empirical study, which makes its points through data and statistical analysis. These two methods follow slightly different models, but both must be guided by the same principles. In all arguments, no matter how they are constructed, the text should clearly show the reasons why the result of the third section is good and should address any objections to or drawbacks of the result.

The example paper by Karaata presents the most important results in the first part of section 4. He first derives the complexity of his algorithm and then shows that the algorithm terminates. If the algorithm were not a polynomial time algorithm, or if it did not terminate in some circumstances, then it might not be useful and worthy of publication. Once the paper has established those properties, it can then argue its value in other cases, as it does in the last part of section 4. Those results are actually some of the more practical aspects of the algorithm that make it appealing. However, these results are subordinate to the first two.

Mathematical Argument. Generally, mathematical argumentation moves from the simple to the complex, as was discussed in section 3 above. Commonly, the paper begins by giving the statement that needs to be proven or established to verify the result. It then begins with simple lemmas and then builds to a general result. As noted in the example paper, the first results in this section usually show that the result is useful, that it will handle the problem that it is attempting to solve. After that, this section usually moves on to special cases that either show the benefits of the result in specific applications or address shortcomings that can affect the result.

However, arguments can also move in a slightly different direction. They can prove a general theorem and then show the specific result as a special case. Both are valid ways of building an argument.

Empirical Arguments. Empirical arguments, arguments based on data and observation, generally follow a simple form, though they regularly vary this form to meet the needs of specific problems. These arguments are essentially two lists. The first list contains the reasons, observations, or deductions that support the validity of the paper's result or conclusion. The second list contains rebuttals or statements that handle any objections to the conclusions.

Both of these lists should be organized in the same fashion, beginning with the strongest points and ending with the weakest. The list of positive statements should begin with the most important evidence that supports the conclusions. The second list, which handles objections, should begin by addressing the most pressing concern about the result and then move to the second greatest concern and so forth until it reaches concerns that are trivial or unimportant.

Sometimes the two lists are interleaved into a single argument that begins with the strongest point in favor of the result and then handles the great objection against the point. From there, it moves to the second most important point and the second objection. The process continues until all the points have been made and all the objections handled. However, it is more common to separate the two, as is seen in the following example, which argues for a new text recognition algorithm.

> In the controlled experiment, our algorithm recognized the target samples 99.99973% of the time and when the samples were restricted to the same texts that had trained the system, it recognized all text presented in front of it. When tested by the field trial, the system correctly identified 96% of the samples, which is 1.4 times more than the standard algorithm that is usually employed in commercial software.

> In the field trial, the algorithm proved to be confused by text that was rotated more than 25 degrees from the standard horizontal, but this can be corrected by multiple application of the algorithm through a cross validation process that identifies the orientation that is closest to horizontal. In this case, the algorithm identified the text correctly 99.99087% of the time at the expense of 11 times more computation. While this is substantially more computation, it is well within the ability of most systems that are applied to this work and only slightly more than most conventional systems.

The greatest problem with the algorithm are the false positives, the times that the system signals that it has correctly identified the text when indeed it has not. As noted in the paper, this occurs only for certain combinations of English letters when they are rotated more than 25 degrees from horizontal. In these cases, the search method will incorrectly map a transformed version of these letters into the dictionary. Based on standard frequency counts, these combinations occur less than 7 times in 10^{18} tests. Furthermore, they are only in certain alphanumeric coding schemes that can be easily avoided in practice by designing the scheme to exclude them.

In this case, the author establishes the validity of the result through the following points:

- 99.99973% success in the controlled experiment
- 96% success in the field trial
- 99.99087% success of modified algorithm

The author then addresses the objections to the method. The first is the additional computational time, which it addresses by claiming that it is an acceptable cost. The second is the number of false positives. The paper argues that such problems rarely occur expect in circumstances that can be designed away.

Section 5: The Recap or Summary or Conclusion

All papers end with a final section that brings the paper to a close. Though this section is often called a summary or a conclusion, it is technically a "recap," which is a short version of the word "recapitulation," which means a restatement of an idea or argument. The final section of a technical paper is *not* a conclusion or a summary, as the conclusions of the work should have been stated in the middle of the paper. It does, however, restate the results and does so in the context of having presented them in full detail in the paper. Hence, the recap tends to give only a quick sketch of the results, a sketch that highlights the main properties of the result and the main reasons that the result is valid.

The recap adds value to the paper by looking carefully at the shortcomings of the result and discussing them as honestly as it can. In this discussion, it is pointing the way toward future research. If the result is not as perfect as desired, the recap should suggest ways of improving the work. If it is not as general as it could be, the recap should note that fact and encourage new research. If it is not efficient enough to be practical, the recap should try to identify ways of improving the work.

In the Karaata paper, the recap is found in section five, which is called "conclusions." Most of the section is devoted to restating the basic result of the paper. As in many cases, this restatement gives the clearest statement of the work, as the author can refer to all the prior work in the paper. At the very end of the paper, the author acknowledges the limits of the work:

> This work is primarily concerned with the placement of resources and/or services in a distributed system. The other aspects of management of resources, such as maintaining the consistency of replicas and the distribution of databases, tables, and services, is outside the scope of this work.

The issues that surround the consistency of replicas and the distribution of databases, tables, and services are the topics that the author sees as being the next step of research.

A Few Additional Guidelines. Recaps and introductions tend to be closely related. Both tend to treat the results of the research broadly and leave the details of the work to the other parts, notably sections 3 and 4. Indeed, many authors will write the first draft of a paper, take the recap paragraphs and move them to the start where they become the introduction and then write a new recap.

Even though the introduction and the recap are related, they are different things. The introduction is supposed to guide the reader into the results of the paper and the recap is supposed to guide them out of it. The introduction avoids some of the details of the research in order to help the reader grasp the nature of the problem before plunging into the depth of the technical work. The recap avoids details because the paper has already presented all the details.

Section 6: Other Elements of a Technical Paper

Abstract. The purpose of the abstract is to allow readers to determine if the paper is of interest to them. It is by nature short and abstract. Unlike the introduction, it makes no effort to guide readers through the background of the field. It presents a brief description of the field of research, the problem being treated, the solution to the problem, and the properties of the solution.

Typically, an abstract has three or four sentences. The first sentence describes the field of research. The second sentence identifies the problem within the field, the standard solution of that problem, and the shortcomings of that solution. The third sentence states the result presented by the paper for addressing that problem and the benefits of that result.

If an abstract has a fourth sentence, that sentence usually deals with additional properties of the result. Karaata's abstract falls neatly into the four sentence model, with a slight modification.

Abstract—In this paper, we present a self-stabilizing algorithm for find-ing clustering of tree networks on a distributed model of computation. Clustering is defined as the covering of the nodes of a network by subtrees such that the intersection of any two subtrees is at most a single node and the difference between the sizes of the largest and the smallest clusters is minimal. The proposed algorithm evenly partitions the network into nearly the same size clusters and places resources and services for each cluster at its clusterhead to minimize the cost of sharing resources and using the services. Due to being self-stabilizing, the algorithm can withstand transient faults and does not require initialization. The paper includes a correctness proof of the algorithm. It concludes with remarks on issues such as open and related problems and the application areas of the algorithm.

In this case, the first sentence not only describes the field, it also states the nature of the result. Such a practice is fine, especially when a paper is presenting a new idea with no prior result or algorithm. The second sentence, "Clustering is defined as the…," de-scribes the problem. In this case, the author makes no reference to prior results, which suggests that there is no other way of handling the problem.

The third sentence, "The proposed algorithm…," describes the result that will be found in the paper. That description requires an additional sentence, "Due to being self-stabilizing…," as it involves a complex property. The next sentence corresponds to the fourth sentence of the model abstract, the one that develops the additional properties. It states that the paper "includes a correctness proof of the algorithm…."

The last sentence in Karaata's abstract is not really needed.

Bibliography. All scientific and technical papers have citations because all of them are part of a literature—they are building upon ideas that have been discovered before and providing a foundation for the research of other workers. All journals and periodicals require that citations for their publication be in a specific style. As these styles vary sub-stantially, we will not discuss the format of citations here. We can only offer the recom-mendation that you should identify the proper style before you submit your paper and make sure that your citations are in that style.

Bibliographies can exhibit one of two different problems: overcitation and undercita-tion. Of the two, undercitation is the more serious. Failing to acknowledge prior work can lead to a charge of plagiarism or intellectual misconduct. In reviewing your work, you should make sure that you give credit to all prior work that contributed to your re-search, including the basic research plan, experimental design, fundamental algorithms, or common results. It is useful to remember that the ideas that are common or funda-mental to you may not be as familiar to an associate editor. It is far, far better to be told by an associate editor that a certain reference is not needed than to be accused of not recognizing prior work.

Overcitation is less of a problem, but it is still a problem, as it forces associate editors to do work that they normally do not do. One does not need to cite all relevant papers back to the fundamental works of Turing, von Neumann, Atkin, and the other pioneers to ensure that the paper acknowledges as all prior work. You need to cite the previous works that define the problem, that provide ideas and materials for your conclusions, and that have produced closely related results.

Again, as undercitation is a more serious problem than overcitation, you should put a citation in your paper if you have any feeling that it might be relevant.

Finally, reviewers regularly suggest papers that should be included in the bibliography. Occasionally, authors balk at these suggestions and claim that the referee is only promoting their own work or the work of their own students. While we cannot comment on the charge that the referees are promoting their own work, we can offer one little bit of advice. The quality of your work will not suffer if it has to include the work of someone else in your bibliography. In the age of cross-references and links, you may even get more readers from the connection. Be brave. Take the citation. Thank the referees.

Keywords. Most publications require that articles be accompanied by keywords. While keywords date to an era that had more primitive ways of storing and retrieving scholarly work, they are still useful as they help categorize research. Some editors use the keywords as a guide in assigning referees.

As with the bibliography, keywords present the challenge of being either too broad or too narrow. If they are too broad, then the keywords can hide the unique contribution of the article and be an obstacle to the results' acceptance in the literature. If they are too narrow, they can bury a paper in a small field and prevent more diverse applications of the idea. Unlike the bibliography, keywords cannot bring charges of plagiarism or author misconduct if they are badly used.

In computer science, most periodicals us the IEEE/ACM classification system, which can be found at http://www.computer.org/portal/web/publications/acmtaxonomy and http://www.acm.org/about/class/1998.

Dealing with Editors and Referees. In scholarly publication, all papers are submitted to the senior editor of a journal, who usually assigns it to an associate editor, who handles the review process. Usually, the associate editor will identify some number of anonymous referees to read the paper and offer their opinions on it. The associate editor will summarize the reports of these reviewers and send them, with a recommendation, to the senior editor. Unless the referees give you their name, you will never know who they are. Often, they will see the paper blind and not know your name.

In the review process, senior editors have to rely on the associate editor and referees beyond their own understanding of the field. They do not get information from other sources. Hence, they tend to stand by the judgments of those referees and associate editors. Therefore, in dealing with the journal, you need to respect the judgment of those referees, even if you are convinced that they are not quite right.

If your article is accepted for publication, you will usually be given the referee reports and be told that you need to follow their suggestions to revise the paper. Very, very few people have papers accepted without a requirement to revise the manuscript. In revising your manuscript, you need to address every issue raised by the referees, though you do not necessarily need to follow their advice. If a referee identifies a certain shortcoming and suggests a strategy for dealing with it, you should acknowledge the issue, even if you decide to address it another way. You should then explain how your approach meets the goals of the referee. Many journals require that you submit a reply to the referees. Even if they do not, it is often a good idea to prepare one, as it forces you to go through the referee reports in a systematic way.

In all cases, reply to the referees with politeness, even if you don't like what they said. Remember, they may be people you would like to have as friends. Or they may be your current friends.

If your paper is rejected and you feel that a referee misread the paper, you can appeal your case to the senior editor. However, most senior editors must follow the guidance of the referee reports and cannot override them. Therefore, you should request that your paper be reviewed a second time. You can politely identify the reasons that suggest that a referee misunderstood your paper. We should note that most periodicals give their senior editors the option to stand by the rejection without a second review, and that very few accept an appeal to the officials who oversee the editors.

Chapter II:
Copyright Law:
What It Is, and How Not
to Break It

C opyright law is complex and somewhat byzantine in its application. It's also the result of the continuous interaction of political, business, economic, philosophical, and practical factors, with the economic ones often taking precedence. However, it all boils down to this: Words and ideas may be more or less intangible, but they're no less ownable than your neighbor's car. You wouldn't dream of stealing your neighbor's car and parking it in your driveway, claiming it as your own. You wouldn't occasionally drive your neighbor's car around town, pointing out to everyone how nice your car is. (At least the authors prefer to believe that you wouldn't think of doing that; if you *are* doing that, please feel free to skip ahead to Section 3 of this chapter, where we'll examine the IEEE's interesting sanctions for plagiarism.)

Assuming that most people wouldn't do that to their neighbor's car, then why would anyone treat someone else's words or unique ideas any differently?

Before we dive in, though, a quick caveat: The IEEE is an international organization, and the IEEE Computer Society includes members and authors from around the world. Here, however, we will focus pretty much exclusively on copyright law as developed and applied in the United States. This is in no way to discount other nations' copyright laws and customs, which are highly relevant both in the individual countries as well as when dealing with the thorny and complex issues of resolving international copyright disputes.

However, it's US copyright law that governs material published by the IEEE, so that will be our topic here. In addition, we'll cover related issues, such as fair use, paraphrasing, educational use, and open access, as well as IEEE's policies regarding plagiarism and author misconduct.

Section 1: Copyright

The Origins and Evolution of US Copyright Law

Like printing itself, the notion of copyright law originated in China. In 1068, less than thirty years after the invention of the world's first moveable-type printing press, the Emperor of the North Song Dynasty issued an order prohibiting the unauthorized reproduction of nine books originally published in 932. One book published shortly after the prohibition, *Stories of the East Capital*, includes the first recorded copyright notice: "Printed by Cheng of Meishan, who applied protection from the superior, any reproduction is prohibited." Nonetheless, other than this imperial order, there was no formal legal or administrative copyright law in China until the mid-twentieth century.[1]

The first formal "copyright law" in the west was the Statute of Queen Anne (1710). It granted a term of 14 years, renewable one time, to book publishers. During that time, publishers enjoyed a complete monopoly on printing, reprinting, and exporting books. (This period is referred to by publishers as "the good old days.")

In the US, copyright was established in the Constitution (Article I, Section 8), in 1789:

> The Congress shall have Power… To promote the Progress of Science and useful Arts, by securing for limited Times to Authors and inventors the exclusive Right to the respective Writings and Discoveries.

---

[1] Wu Shulin, "The Conditions of the Judicial and Administrative Protection of Copyright in China," *Duke Journal of Comparative and International Law*, vol. 9, no. 1, 1998, pp. 241–248, http://scholarship.law.duke.edu/djcil/vol9/iss1/12.

Copyright is codified in the Copyright Act in the US Code (the Constitution provides a general principle; the Code fleshes it out, explains it, and renders that principle enforceable).

Two important phrases to remember from the US Constitution are "promote progress" and "limited times."

Supreme Court Justice Sandra Day O'Connor had this to say about copyright law, and it's as good a summing-up as any of the role of copyright—or the role it ought to have—in a democratic and capitalist society:

> The primary objective of copyright is not to reward the labor of authors, but "[t]o promote the Progress of Science and useful Arts." To this end, copyright assures authors the right to their original expression, but encourages others to build freely upon the ideas and information conveyed by a work. This result is neither unfair nor unfortunate. It is the means by which copyright advances the progress of science and art.[2]

As we'll see, politics and economics are not entirely irrelevant to a discussion of copyright law.

Current US Copyright Law

Lawyers like analogies, and intellectual property lawyers are no different. The set of legal rights enjoyed by a copyright holder can be thought of as a bundle of sticks. The five rights of a copyright holder are the rights to publish, adapt, perform, reproduce, and display the copyrighted work.

These rights belong to the author alone from the moment a work is created in tangible form (more on that shortly). However, the author is not required to hold on to those rights. Because they belong to authors, they are free to do with them as they please.

The copyright holder can give or sell any or all of those sticks to others (in a standard publishing agreement, for instance, authors generally assign their rights to the publisher). They can do so with any restrictions they like, or with none at all. For instance, an author could totally sign over copyright to a publisher forever, or grant the publisher an exclusive right to reproduce and distribute the work for a limited period of time.

In other words, if I write a novel, I have the right to keep my novel locked in a desk drawer, or to permit a publisher to print and sell it, or to allow someone to adapt it as a musical on Broadway or rewrite it as a poem or produce it as a 3-D movie. And I'm the

---

2   Feist Publications, Inc. v. Rural Telephone Service Co., *US Reports*, vol. 499, 1991, pp. 340, 349 (Justice Sandra Day O'Connor).

only one who has that right, unless I've given someone else the "stick" that lets them make that decision for me.

As long as the author retains those "sticks," he or she is the only person on the planet who has the legal right to publish a work, adapt or make changes to it, perform or present it publicly, make copies of it, or show it to others. *No one else has the legal right to do any of those things.*

When Does Copyright Begin?

It's a common misperception that a work is protected by US copyright law only if it is registered with the Library of Congress. In 1989, the United States became a signatory to the Berne Convention, an international agreement in which more than 160 countries agreed to abide by some fundamental principles protecting intellectual property. The Convention imposes certain minimal standards, such as the automatic creation of copyright and a baseline copyright term of the author's life plus fifty years.[3]

So while registering a copyright automatically provides a number of specific legal protections and access to more severe remedies against infringers, registration is not required for a work to be legally protected and for an author to hold those sticks.

Under US law, an author's copyright is created at the moment a work takes tangible form. "Tangible form" includes nonprint media, so a Word file on a computer constitutes sufficiently tangible form for a work to be copyrighted. On the other hand, merely thinking about a work or talking about it to others does not create a copyright.

As soon as a work has a tangible form, then, an author is entitled to let the world know that he or she is holding all the sticks by displaying a © symbol next to his or her name, along with the year in which the work was created. That year is important, because it starts a count-down to the author's loss of copyright protections—as we'll see next.

How Long Does Copyright Last?

Over the centuries, the length of time a work is legally protected by copyright, including available renewals, has grown longer.

This is a good example of how economics and law are not always unrelated. As publishing became a bigger and more widespread business over the centuries, with money to be made or lost based on the publishers' and authors' exclusive rights to the content being sold, the legal structure that protects artists and publishers responded accordingly.

Here are the current terms for copyright protection under the Sonny Bono Copyright Extension Act of 1998 (also referred to as the Millennium Copyright Act, and officially "An Act to Amend the Provisions of Title 17, United States Code, with Respect to the Duration of Copyright"):

---

3   WIPO, "Berne Convention for the Protection of Literary and Artistic Works," as amended, www.wipo.int/treaties/en/ip/berne/trtdocs_wo001.html.

- Works created after 1/1/1978: author's life + 70 years
- Works created but not published before 1/1/1978: author's life + 70 years; if the work was published by 12/31/2002 then the term will not expire before 12/31/2047
- Pre-1978 works still in original or renewal copyright term: 95 years from original copyright date.

The most recent revision to copyright law, the Millennium Copyright Act, was championed in large part by the Disney corporation, which feared the loss of exclusive rights to one of its highly profitable properties. Interestingly, the copyright on Mickey Mouse was due to expire, under the old law, in 2003 (as an artwork, Mickey is copyrightable). Donald Duck, Goofy, and Pluto would have entered the public domain in 2009. As you might imagine, this could create a significant economic hardship for the Walt Disney Company, which aggressively protects its copyrights. Disney spearheaded a copyright-reform effort that brought pressure on lawmakers to adjust the "unfair" copyright terms then in effect.[4]

Like we've said, the law works to protect the interests of the copyright holder, who stands to experience considerable economic hardship once control of the property is lost to the public. Or as the US Supreme Court put it, "[T]o negate fair use one need only show that if the challenged use 'should become widespread, it would adversely affect the potential market for the copyrighted work.'"[5]

Section 2: Fair Use

It's already been stated that no one other than the author has any rights to a copyrighted work unless the author has specifically transferred those rights. In this section, we'll explore some of the loopholes—or, rather, "exceptions"—to this general rule that permit the free use of copyrighted material *without* the copyright holder's permission. This should not be considered a guide to "how to plagiarize," but rather as a broad summary of how to *avoid* plagiarizing.

"Fair use" is a legal term that refers to Section 107 of the Copyright Act.[6] The Copy-

---

4   An interesting analysis of this battle is provided by Chris Sprigman, "The Mouse That Ate the Public Domain: Disney, the Copyright Term Extension Act, and *Eldred v. Ashcroft*," http://writ.news.findlaw.com/commentary/20020305_sprigman.html.

5   Harper & Row, Publishers, Inc., and the Reader's Digest Association, Inc., Petitioners v. Nation Enterprises and the Nation Associates, Inc., *US Reports*, vol. 471, 1985, p. 568, quoting Sony Corp. v. Universal City Studios, *US Reports*, vol. 464, 1984, p. 451 in Princeton Univ. Press et al. v. Michigan Document Services, *Federal Reporter*, 3rd Series, vol. 99, 1996, p. 1381 (US Court of Appeals for the Sixth Circuit).

6   US Code, Title 17, section 101 et seq. as amended.

right Act defines the rights of the copyright holder and how copyright is protected against infringement. Section 107 describes the "fair use exception"—how, under certain limited circumstances, individuals are free to use copyrighted materials word for word without the permission of the copyright holder (the author or publisher).

Prior to the 1976 revision of the Copyright Act, "fair use" had evolved as a court-made doctrine. Congress essentially codified the common law principles that had developed over time into a statutory "fair use" law. Basically, the four factors governing fair use are:

1. The **purpose and character** of the use;
2. The **nature** of the work in which the material is used;
3. **How much** of the copyrighted work is being used (and how substantive it is); and
4. The **economic effect** of the use on the copyrighted work's market.

It's important to remember that the purpose of copyright is to protect both an author's artistic interest in the work and, more importantly from a legal perspective, his or her economic interest.[7] If someone lifts large portions of a copyrighted work for use elsewhere, without the author's permission, it can adversely impact the author's revenue stream.

For instance, if the Computer Society were to decide it would be beneficial to its members to post a chapter a month of the *Twilight* books on computer.org, many computer professionals would have the opportunity to read all about pale, moody teenage vampires without actually paying for it. That would deprive Stephenie Meyer of her royalties and would put a dent in Little, Brown Books for Young Readers' revenue stream. Neither Ms. Meyer nor Little, Brown would be pleased, and they would be entitled to seek legal action and compensation from the Computer Society. (Of course, the Computer Society would never do such a thing.)

Exceptions to Every Exception

But that's an example of something that is *not* a fair use. What *would* qualify for the exception? Here are some random examples, based on what US courts have held over the years:

- Quotation of excerpts in a review or criticism of the work, with citation.
- Quotation of short passages in a scholarly or technical work for the purpose of clarifying the author's observations, with full citation to the source.

---

7   "[T]o negate fair use… one need only show that if the challenged use 'should become widespread, it would adversely affect the potential market for the copyrighted work.'" Harper & Row, Publishers, Inc., and the Reader's Digest Association, Inc., Petitioners v. Nation Enterprises and the Nation Associates, Inc., *US Reports*, vol. 471, 1985, p. 568, quoting Sony Corp. v. Universal City Studios, *US Reports*, vol. 464, 1984, p. 451 in Princeton Univ. Press et al. v. Michigan Document Services, *Federal Reporter*, 3rd Series, vol. 99, 1996, p. 1381 (US Court of Appeals for the Sixth Circuit).

- Parody that includes some content of the parodied work.
- Summary in a news article or report.
- Reproduction by a library to replace portions of a damaged copy.
- Reproduction in legislative or judicial proceedings or reports.
- Incidental reproduction in a broadcast of a work located in the scene of an event being reported.
- Last-minute reproduction by a teacher, where there is insufficient time to obtain permission.

In other words, if your use of the material is similar to any of these examples, you're probably OK.

However, it's important to note that "fair use" is not a right; it's an "affirmative defense" to an accusation of copyright infringement. That is, the person accused of infringement would acknowledge that the copyrighted material was used without permission, but would defend that violation by claiming that the use fell into one of the standard permissible "fair use" exceptions.

Other affirmative defenses to copyright infringement include:

- **Minor use**: A claim that the portion of the copyrighted work that was used is so small and insignificant that it's a waste of everyone's time to be in court at all. (For instance, Professor Clausula sues Professor Stantibus for infringement, based on Stantibus' repeated use of the word "algorithm" in her scholarly paper. Clausula's claim that the word appeared in his earlier-published paper will likely not stand, since the common one-word portion that is repeated is extremely minor and inconsequential. On the other hand, if Stantibus copied six pages of Clausula's seven-page paper, her claim that it was a minor use because she only copied the six pages one time in her eighty-page paper would be unlikely to convince a judge.)
- **Expired copyright**: A claim that the original copyright has expired, and that the work is no longer legally protected. This is tricky, and depends on a careful and accurate parsing of the interaction of various copyright laws. (For example, Margaret Mitchell's novel *Gone with the Wind* was published in 1936. Under the copyright law existing at that time, *GWTW* would have been in the public domain—that is, free for all to use—in 1964. By 1964, the copyright law permitted renewal, and an additional 28 years were added to the copyright term. In 1978, Congress changed the renewal term from 28 to 67 years, so the novel's copyright was extended for an additional 39 years and won't expire until 2031. The most recent copyright law reform didn't have any impact on this work because of the date of its original publication.)
- **Transformative use**: A use is considered "transformative" if the copyrighted work was used to help create something completely new and unrelated to the original. (For example, a photographer sued artist Jeff Koons for depicting part of one of her photographs, which had been commissioned for a magazine advertisement, in a painting. Koons won, and the court observed:

The test for whether [a work] is "transformative," then, is whether it "merely supersedes the objects of the original creation, or instead adds something new, with a further purpose or different character. [This test] perfectly describes Koons's adaptation of "Silk Sandals": the use of a fashion photograph created for publication in a glossy American "lifestyles" magazine—with changes of its colors, the background against which it is portrayed, the medium, the size of the objects pictured, the objects' details and, crucially, their entirely different purpose and meaning—as part of a massive painting commissioned for exhibition in a German art-gallery space. We therefore conclude that the use in question was transformative.[8])

"Educational Use"

While the "educational use" exception is not entirely relevant to a work on how to avoid copyright violations in scholarly papers, since many scholarly papers are written by educators, this seems like a convenient place to mention it.

While education is most certainly a very noble endeavor, 17 USC § 107 still provides *very* strict rules for invoking the "educational use" exception to the Copyright Act. The "educational use" exception is an extension of the fair use rules designed to accommodate the special needs and circumstances of educators. Many educators, however, blithely assume that just invoking the magic words "educational use" gives them carte blanche to use copyrighted materials in their classrooms willy-nilly.

That assumption is totally untrue.

Here's the truth: An instructor may make multiple copies of a copyrighted work for classroom use *only* if *all* of the following requirements are met:

1. The copying meets the test of **brevity**
2. The copying meets the test of **spontaneity**
3. No more than **nine instances** of multiple copying take place during a term, and only a limited number of copies are made from the works of any one author or from any one collective work
4. Each copy contains a **notice of copyright**
5. The copying does not **substitute** for the purchase of "books, publishers' reprints or periodicals" and
6. The student is **not charged** any more than the actual cost of copying.

"Brevity" has different meanings, depending on the original content. For prose, "brief" means either (i) a complete article, short story, or essay of less than 2,500 words,

---

8   Blanch v. Koons, No. 05-6433, *Federal Reporter,* 3rd Series, vol. 467, 2006, p. 244 (US Court of Appeals for the Second Circuit).

or (ii) an excerpt from any prose work of not more than 1,000 words or 10 percent of the total work, whichever is less. A complete short poem may be reprinted, if it's less than 250 words, or up to 250 words from a longer poem. Regarding illustrations, the brevity exception applies only to one chart, graph, diagram, drawing, cartoon, or picture per book or periodical (i.e., one editorial cartoon from today's *New York Times*, but *not* an editorial cartoon and a "Doonesbury").

The rule for spontaneity is even simpler: one instructor at the last minute. The Act says, "The inspiration and decision to use the work and the moment of its use for maximum teaching effectiveness [must be] so close in time that it would be unreasonable to expect a timely reply to a request for permission," and the copying must be at the instance and inspiration of an individual teacher.[9]

That means you have to wake up in the morning two hours before class and be instantaneously inspired by, say, an editorial cartoon you see in the digital edition of your morning paper. You're allowed to download, print, and photocopy the cartoon for distribution in that morning's class, because there's not enough time to contact the paper for permission. What you can't do is then distribute that cartoon to classes every semester, unless you've received permission.

The exception strictly limits how often inspiration can strike an educator (nine) and naturally requires that any materials used must include proper attribution and a statement of copyright. Finally, educators can't make this into a cottage industry by charging students more than it cost to reproduce the materials.[10]

Section 3: The IEEE and Copyright

The IEEE's copyright policy is straightforward. Basically, if a paper is accepted for publication by the IEEE, then the author is expected to execute a transfer of copyright. IEEE holds the full and exclusive copyright to all the contents of the Xplore digital library.

That's not to say that the author gives up everything. The IEEE policy also permits authors to publish their papers *in the form in which they were originally accepted* on their

---

9   US Code, Title 17, section 107, reproduced at Cornell Law School Legal Information Institute, www4. law.cornell.edu/uscode/17/107.notes.html.

10   It can be very challenging to determine who holds the copyright to material like editorial cartoons or television programs, and then it's often additionally frustrating trying to find the right person to approach. Corporations, syndicates, and film studios are often not particularly willing to grant permission even to educators. Further, they tend to charge impressive (some might say excessive) fees even for innocent, non-profit, educational, or scholarly uses. Sadly, this does not create an incentive for educators or researchers to comply with what many see as restrictive, outdated, short-sighted, anti-scholarly and anti-creative laws and attitudes. Nonetheless, those are the rules at least for now, and failing to comply puts individuals and their institutions at risk.

personal or institution's websites. "The form in which they were originally accepted" means the non-edited, non-proofread, non-typeset version that finally appears in a magazine, conference proceedings, or transaction. A notice of IEEE copyright and a full citation of the paper must appear on the first page of the posting.

There are, then, two definitions of a paper that you need to keep in mind when dealing with IEEE copyright and usage questions:

1. The "accepted version." The accepted version is the final manuscript accepted by an IEEE publication, as the author submitted it to IEEE.

2. The "published version." This is the way the article or paper actually appears as published in print or digital format, and as it is archived in Xplore.

The published version has gone through the IEEE's internal publishing process and has been copyedited for adherence to IEEE style, formatted and designed to conform to the specifications of a particular publication, and generally "prettied up" for distribution to the public. In business terms, IEEE has "added value" to the submitted version: contributed the icing on the cake, as it were, or the wrapping paper on the gift. IEEE's editorial and production staff have not altered the content of the author's work, but rather made it commercially appropriate.

Instructors are permitted to use all or part of any work posted to Xplore in their classes, as long as they use the "accepted version" rather than the published version. And IEEE will accept content that was first posted on an institutional site, for instance, as long as the author is able to transfer copyright to IEEE. So, for example, if a professor posts an original scholarly work on her university's website for student use, she may later submit the work to IEEE for publication if her university allows her to retain copyright. And if the author has posted an unpublished manuscript on a public website, the author is required to attach an IEEE copyright notice to the work once it is accepted for publication by IEEE; the author can leave the paper posted on the site, however.

An appropriate citation for a paper published by IEEE but posted elsewhere will include a full standard citation, including the Digital Object Identifier (DOI) or a link to the paper's abstract.[11]

Open Access

The IEEE offers authors three Open Access (OA) models: *IEEE Access* is a "megajournal," a 100% OA publication with a scope including all areas of the IEEE's field of interest and expedited peer review. Topical OA journals, such as the Computer Society's *Transactions on Emerging Topics in Computing* (*TETC*), are more focused in scope. And *all* IEEE journals offer a hybrid option under which authors may elect that their individual paper be of-

---

11 For more on the IEEE policy, interested authors can go here: http://www.ieee.org/publications_standards/publications/rights/rights_policies.html.

fered as OA, whether or not the journal in which it appears is OA. All three options require that authors pay a publishing fee, but authors retain their copyrights.

Plagiarism

Sadly, not everyone recognizes the simple rule that if it's not yours, don't pretend it is. There are no doubt a variety of socio-cultural reasons underlying the rising rate of plagiarism across all scientific, technical, and academic disciplines. The Internet in particular, by providing widespread access to copyrighted material (a good thing), has made it alarmingly simple for individuals to "borrow" content from others without attribution (a bad thing). Because plagiarism is an increasingly widespread and serious problem, the IEEE and IEEE Computer Society have worked together to establish definitions, investigative procedures, and penalties for the misuse of content that was created by others.

The IEEE Computer Society (CS) follows section 8.2.4 of the IEEE Publication Services and Products Board (PSPB) Operations Manual for the investigation of all author misconduct complaints. That document is the official statement of how the IEEE defines and deals with issues of plagiarism. Here, we will attempt to explain that approach and give an example of how it works. However, the IEEE documents remain the final authority on plagiarism, author misconduct, and the procedures for handling complaints about such issues.

Defining Plagiarism

The IEEE's definition of plagiarism, found in section 8.2.1 (B)(7) of the PSPB Manual, is "the use of someone else's prior ideas, processes, results, or words without explicitly acknowledging the original author and source."

The IEEE identifies five degrees (or "levels") of plagiarism. Those levels are explained in the manual. According to the IEEE, the five levels of plagiarism are:

- Level One: The uncredited verbatim copying of a full paper, or the verbatim copying of a major portion (greater than half of the original paper).
- Level Two: The uncredited verbatim copying of a large portion (less than half of the original paper).
- Level Three: The uncredited verbatim copying of individual elements (e.g., paragraphs, sentences, figures).
- Level Four: The uncredited improper paraphrasing of pages or paragraphs
- Level Five: The credited verbatim copying of a major portion of a paper without clear delineation (e.g., quotes or indents).

Let's look at a quick example to illustrate what these levels mean. Below is a brief passage from page 6 of David Alan Grier's book, *When Computers Were Human* (Princeton

University Press, 2005). For our purposes here, we'll pretend that this passage is an entire paper. Also, in each of the examples of plagiarism, we'll clearly identify the original material from Dr. Grier's book—something a plagiarist would never do.

> The story of organized scientific computation shares three themes with the history of labor and the history of factories: the division of labor, the idea of mass production, and the development of professional managers. All of these themes emerge in the first organized computing groups of the eighteenth century and reappear in new forms as the story develops. All three were identified by Charles Babbage in the 1820s, when he was considering problems of computation.

An instance of Level One plagiarism might look like this:

> The development of computers in science and technology has a long and robust history. *The story of organized scientific computation shares three themes with the history of labor and the history of factories: the division of labor, the idea of mass production, and the development of professional managers. All of these themes emerge in the first organized computing groups of the eighteenth century and reappear in new forms as the story develops. All three were identified by Charles Babbage in the 1820s, when he was considering problems of computation.* All three continue to this day.

In this example, the entire "paper" is incorporated word for word into the plagiarizing author's work, without attributing the source or indicating that it's quoted material. The author is simply blatantly representing Dr. Grier's work as her own.

A Level Two instance of plagiarism could look like this:

> The development of computers in science and technology has a long and robust history. It *shares three themes with the history of labor and the history of factories: the division of labor, the idea of mass production, and the development of professional managers. All of these themes emerge in the first organized computing groups of the eighteenth century.* They were initially recognized by the earliest pioneers of scientific computing, and all three continue to this day.

Once again, the author is taking credit for Dr. Grier's words (in effect, stealing the content). Here, however, it's not the full "paper" that is taken, but a *portion* of it.

Level Three plagiarism is slightly more complex:

The development of computers in science and technology has a long and robust history. *The story of organized scientific computation has three themes in common with the history of labor and factories: the idea of mass production, the division of labor, and professional managers. Charles Babbage identified all three in the 1820s, when he was working on problems of computation.* All three continue to this day.

In this example, the plagiarizing author has woven the original paper into his own, taking certain passages verbatim, slightly modifying others (but leaving them identifiably Grier's words), and failing to indicate that any of the content is derived from another source.

Level Four is a somewhat lesser offense, in which the plagiarizing author paraphrases Dr. Grier's words, but retains his essential ideas, again without citing the original source. Some authors mistakenly believe that they can avoid charges of plagiarism by paraphrasing the original work, or putting it "in their own words." However, plagiarism is committed whenever one author uses another's *ideas, processes, results, or words* without attribution.

In Other Words: The Pitfalls of Paraphrasing

Paraphrasing is tricky, because while you're avoiding the direct use of someone else's words, you're retaining their ideas (so you still need to cite your source). Paraphrasing also runs the risk of changing the original author's intended meaning. Particularly if you're providing attribution (which you always should), poor paraphrasing is basically putting words in someone else's mouth.

Charles Caleb Colton (1780–1832) was an English wine merchant, sportsman, writer, cleric, eccentric, and wit whose primary legacy to the world is this phrase:

> Imitation is the sincerest form of flattery.

Here's an *appropriate* paraphrase of Colton's maxim:

> According to Colton, the most sincere compliment is to imitate someone's actions. (C.C. Colton, *Lacon*, 1820.)

But here is an example of *faulty paraphrasing*, in which the meaning of the original is altered by the well-intentioned abuse of the thesaurus:

> According to Colton, if you really want to say something nice to someone, the best thing to do is to copy everything *they* say. (C.C. Colton, *Lacon*, 1820.)

This paraphrase is faulty because it makes Colton's saying sound like a casting call for impressionists. Instead of emphasizing Colton's underlying meaning (imitation demonstrates sincere admiration) this passage tries so hard to avoid using Colton's words that it winds up suggesting that mimicking someone is a good thing. As anyone who has encountered a six-year-old who has just decided to repeat everything you say for several hours can attest, this is untrue.

Going back to Grier's "paper," however, a paraphrase would fall under the IEEE's definition of Level Four plagiarism if it looked like this:

> The development of computers in science and technology has a long and robust history that has three things in common with the development of labor and factories:
>
> 1. mass production,
> 2. division of labor, and
> 3. professional managers.
>
> These elements emerged as part of the first groups of researchers engaged in computing activities in the 18th century, and continue—although evolving—as the story of computers goes on. In the 1820s, Charles Babbage was the first person to discover them when he was thinking about issues involving computations. All three continue to this day.

Here the plagiarizing author has just reworded and reformatted Grier's basic ideas without any attribution. Also, in paraphrasing Grier's sentence about Babbage "first noticing" the three items, the new version suggests that Babbage "discovered" them, which may or may not be what Grier meant.

Finally, Level Five plagiarism is really just sloppy scholarship:

> The development of computers in science and technology has a long and robust history. *The story of organized scientific computation shares three themes with the history of labor and the history of factories: the division of labor, the idea of mass production, and the development of professional managers. All of these themes emerge in the first organized computing groups of the eighteenth century and reappear in new forms as the story develops. All three were identified by Charles Babbage in the 1820s, when he was considering problems of computation.* Karl Marx, Babbage's contemporary, explored the relationship between technology and labor as an economic issue, what he referred to as a struggle between worker and the machines that were designed to eliminate the worker as a unit of labor. All three of those

components, and the economic tension between machine and human, continue to this day. (D.A. Grier, *When Computers Were Human*, Princeton University Press, 2005.)

Here, the plagiarizing author—whose interests seem to lie more in economics than computer science—properly credits Grier's work, but fails to indicate exactly where the content is quoted. In this case, that failure makes Grier appear to be more of an economic historian than one whose primary interest is technology. (It also appears that this author may have plagiarized the non-Grier material as well: no source is cited and it comes directly from Nick Dyer-Withford's *Cyber Marx: Cycles and Circuits of Struggle in High-Technology Capitalism*, Board of Trustees of the University of Illinois, 1999).

Plagiarism: Investigation
The IEEE and the IEEE Computer Society take charges of plagiarism very seriously. An allegation of plagiarism immediately triggers an investigation, and there are specific procedures that are followed to investigate allegations of plagiarism.

Complaints may be received by either the IEEE Computer Society (CS) or the IEEE Intellectual Property Rights office (IPR). The CS investigates all cases of author misconduct whether previously published or recently submitted and still under review.

All misconduct receives the same rigorous investigation whether it was published in the past and discovered in Xplore or just recently submitted to a CS-sponsored conference, magazine, or journal. The paper in question may also be published by or submitted to an outside publisher, plagiarizing a CS paper.

All papers under investigation are first entered into the Ithenticate CrossCheck plagiarism detection system, and the report generated by the system is used during the investigation.

The CS works directly with the IPR office during the investigation. The IPR office prepares marked copies of the accused and original papers and issues a summary report of the findings. The recommended decision is then presented to an investigation committee. The investigation committee consists of two or more senior-level CS volunteers.

An author who is accused of plagiarism is asked for a response and is provided the marked copies from the IPR office. All co-authors with available email addresses are included in the request for a response. If no email address can be found for an author, the investigation goes forward without a response. A deadline is given for the accused author to respond to the allegation. If no response is received by the due date, the investigation proceeds.

When the accused author responds (or the due date for response has passed) all information is sent to the investigation committee for a decision. Decisions of level one or two plagiarism are sent to the PSPB Vice President for approval. The decision is then sent to the offending author and complainant. The IPR office will also add the offending author name and time of publication ban to the prohibited author list. A notice of violation of publication principles is added to the paper in Xplore at the same time that the notification is sent to the authors.

The next section outlines the specific penalties imposed for the various levels of plagiarism.

Plagiarism: Penalties

The specific penalties for committing plagiarism vary depending on the degree (level) of the offense. The IEEE and IEEE Computer Society corrective actions for plagiarism are as follows:

Level One:
- Publication of a Notice of Violation of Publication Principles in IEEE Xplore and the CSDL in the appropriate IEEE publication;
- Prohibition of publication in all IEEE-copyrighted publications by the offending authors(s) for 3 to 5 years;
- Rejection and return of all papers by the author(s) that are currently in review or in any IEEE publication queue (papers may be re-submitted after prohibition term has expired).

Level Two:
- Publication of a Notice of Violation of Publication Principles in IEEE Xplore and the CSDL in the appropriate IEEE publication;
- Prohibition of publication in all IEEE-copyrighted publications by the offending authors(s) for 1 to 3 years;
- Rejection and return of all papers by the author(s) that are currently in review or in any IEEE publication queue (papers may be re-submitted after prohibition term has expired).

Level Three:
Offending author(s) prepare and submit an apology letter to the plagiarized author(s) and to the editor of the publication where the plagiarized work was submitted. Failure to comply with the corrective actions may result in additional, more severe corrective actions, including prohibition from publication for 1 to 2 years.

Level Four:
Offending author(s) prepare and submit an apology letter to the plagiarized author(s) and publication editor. Failure to comply with the corrective actions may result in additional, more severe corrective actions, including prohibition from publication for 1 to 2 years.

Level Five:

Offending author(s) prepare and submit an apology letter to the original author(s) and publication editor, and prepare correction or retraction and submit this document to the editor for publication. Failure to comply with the corrective actions may result in additional, more severe corrective actions, including prohibition from publication for 1 to 2 years.

Conclusions

As promised, we've covered quite a bit of territory in a relatively short space: the Song Dynasty, the Berne Convention, the US Constitution, the US Code, some case law, open access, fair use, and the IEEE Policy & Procedures Manual. In the end, where does it all leave us? Right back where we started:

> Words and ideas may be more or less intangible, but they're no less ownable than your neighbor's car.

That's not a difficult concept, really. With some few exceptions, with some limitations on purpose and quantity, and wrapped up in the intricate and impenetrable language of judicial decisions, codified law, international treaty, and technical society rules and bylaws, it's about as basic as something your mother might have said:

If it's not yours, don't pretend it is.

Closer to home, scholarly research and writing is a process that builds upon itself, and is inherently collaborative, derivative, and transformative. The work of one researcher sparks ideas in another, original research is replicated, ideas are tweaked and adjusted and expanded or contracted as the sphere of knowledge expands. Much of what is written in our community relies heavily on what's been written before. That's all very fancy-sounding, but it means it's more important here than almost anywhere that others' work be respected, credited, cited, and used in a way that conforms with not only the law of copyright, but with its spirit: again, in plain English, if it's not yours, don't pretend it is.

9455505R00028

Printed in Great Britain
by Amazon.co.uk, Ltd.,
Marston Gate.